SUPER CHALLENGE

SUPER
CHALLENGE

WORLD BOOK, INC.

CHICAGO LONDON SYDNEY TORONTO

World Book, Inc.
525 W. Monroe
Chicago, IL 60661
U.S.A.

Editor: Melissa Tucker
Design: Lisa Buckley
Cover design: Design 5

Library of Congress Cataloging-in-Publication Data

Super challenge.
        p.   cm. -- (World Book's mind benders)
      Summary: A collection of puzzles emphasizing creative thinking and
   deduction, some of which involve numbers.
      ISBN 0-7166-4108-9 (softcover)
      1. Mathematical recreations--Juvenile literature.   [1. Puzzles.
   2. Mathematical recreations.]   I. World Book, Inc.   II. Series.
   QA95.S737 1997
   793.7'4--dc21   97-6017

**For information on other World Book products, call 1-800-255-1750, X2238,
or visit us at our Web site at http://www.worldbook.com**

Printed in Singapore.

1 2 3 4 5 99 98 97

Introduction

The puzzles in this book have one thing in common—they are designed to make you think carefully.

In some of the puzzles, you have to decide what kind of question someone might ask to get a certain answer. There aren't really any clues in this kind of puzzle. You have to ask yourself questions and see what the answers might be. Sooner or later, you'll get an idea of what the right sort of question is.

One or two of the puzzles are really based on common sense. You have to ask yourself, "How would *I* do this?"

Don't give up too easily. When you solve one of these puzzles, you can be really proud.

The people of Tuffleheim

Long ago, a traveler walking along a road met a man coming the other way. "Hello, friend," called the man. "Where are you heading?"

"I'm going to the town of Tuffleheim," said the traveler.

"Oh, you'll need to be careful, there," warned the other man. "That town has been enchanted by a wicked magician! Half the people can tell only the truth, and the other half can tell only lies. You never know whom to believe in Tuffleheim."

"I'll be careful," said the traveler.

An hour later, he reached Tuffleheim. It was getting late, and he was tired and hungry. He wanted to find a good inn where he could have a nice dinner and a comfortable room in which to spend the night. In any other town, he would have simply asked someone for the name of the best inn. But he couldn't do that, here. For, if the person he asked should happen to be one of the liars, he would send the traveler to the *worst* inn.

The traveler realized he would have to first find out whether anyone he talked to was a liar or a truth-teller. He went up to the first Tuffleheimer he saw and asked her a question. What question could the traveler have asked that would let him know, at once, whether a person was a liar or a truth-teller?

(ANSWER ON PAGE 29)

How many carnations?

Wilfred McDoodle was color blind. Red and green both looked gray to him.

Wilfred got a job working for a florist. One day, when he was all alone in the store, a lady came in and wanted to buy either two red or two green carnations. She didn't care which color they were so long as they were both the same.

Wilfred knew there were a dozen red carnations and a dozen green carnations all mixed together in the refrigerator in the backroom of the store. But the red and green carnations all looked the same to him. Would he have to bring all two dozen carnations so the lady could pick out two of the same color?

What is the smallest number of carnations Wilfred would have to bring back to be sure of having two of the same color?

(ANSWER ON PAGE 29)

The hungry rabbit

A rabbit was making a long journey across a plain.
There was absolutely nothing on the plain for her to
eat. By the time she was halfway across, she was starving.
She was much too weak to keep going.

However, halfway across the plain there was a small
house. The owner of the house was a cabbage farmer.
In the yard next to the house was a big pile of cabbages.
They were surrounded by a square fence.

The rabbit had to eat or she would soon die. She was
thin enough to squeeze through the fence and get to the
cabbages. But if she went into the enclosure and ate as
much as she needed, she would be too fat to get back
out. The farmer would probably kill her. And the
cabbages were all too big to be pushed through the
fence to the outside.

What can the rabbit do?

(ANSWER ON PAGE 29)

The outlaws and the marshals

Sheriff Dalton of Dry Gulch heard that two U.S. marshals were searching for some outlaws near his town. He decided to ride out and see if he could help them.

After a time, he saw a campfire with four men around it. Riding closer, he saw that two of the men were tied up and two were untied.

"Are you the marshals?" he called to the untied men.

One of the men said something, but Sheriff

Dalton couldn't hear him. "What did you say?" he called.

The second man answered. "He said that we're the marshals."

Suddenly, one of the tied-up men began to struggle. "That's not true!" he yelled. "*We're* the marshals! We set out to capture these two, but they ambushed us! They're the outlaws."

Sheriff Dalton knew that the marshals would always tell the truth. He also knew that the outlaws would always tell lies. How could he know, from what the men said, which ones were really the marshals and which ones were the outlaws?

(ANSWER ON PAGE 29)

Caspuccio the bandit

In the kingdom of Cappedonia, there was a daring bandit named Caspuccio. But he robbed only barons, earls, and wealthy nobles. And he gave much of the money he stole to poor people who needed it. So, most common people liked him. Of course, the barons and earls hated him.

Finally, Caspuccio was captured. He was taken before King Bombaso, for trial. All the barons and earls came to the trial. So did many common people.

"So, this is the famous robber, Caspuccio," said the king. "What shall we do with him?"

"Hang him!" yelled all the barons.

"Cut off his head!" shouted the earls.

"Set him free!" cried the common people.

"Well, I must be fair about this," said the king. "I'll give everyone a chance to have their way. Caspuccio, I want you to make a statement. In other words, *tell* me something. If what you say is true, you'll be hanged, as the barons want. If what you say is false, you'll have your head cut off, as the earls want. But if I can't decide if it's true or false, you'll go free, as the common people want."

Caspuccio knew he didn't have much of a chance. After all, every statement is either true or false! The king could say that what Caspuccio said was either true or false.

Caspuccio thought for a few moments. Then, a smile came over his face. He spoke seven words. And after thinking about what he said for a while, the king set him free!

What did Caspuccio say?

(ANSWER ON PAGE 29)

The magical beanies

There are two girls, named Kim and Hannah, and three boys, named Mark, Frank, and John. They are all wearing beanie caps. Some of the caps are red and some are white. Each child can see only the other children's caps.

The beanies are magical. Anyone wearing a white beanie can tell only the truth. Anyone wearing a red beanie can tell only a lie.

Frank says, "I see three red beanies and two white ones."

Mark says, "I see four white beanies."

Hannah says, "I see three red beanies and one white one."

Kim says, "I see one red and three whites."

John says, "I see what Hannah sees."

From what the children said, can you tell the color of the beanie each one was wearing?

(ANSWER ON PAGE 30)

The three musicians

Long ago, in China, three musicians were traveling to the city of Peking. They had been hired to perform at a wedding there.

The three came to a wide river. They were shocked to see that the bridge had collapsed. There was no way to get across the river. The musicians knew that the nearest bridge was far down the river, nearly a day's walk away. But they had to be in Peking in a few hours.

Suddenly, they saw two children rowing a boat down the river. The musicians called to the children, offering to pay them to use the boat. The children rowed to the shore.

However, the boat was *very* small. It could hold only two children or one grown-up. This was a serious problem. If one of the musicians took the boat across the river, there would be no way to send the boat back. A grown-up couldn't take a child and then send the child back with the boat—for a grown-up and child couldn't both get into the boat at the same time.

But one of the children had an idea. And in less than an hour, all three musicians were across the river. How did they do it?

(ANSWER ON PAGE 30)

The brave clown

The guard at the bridge looked surprised. Then he smiled. Walking down the road came a chubby clown, juggling three colored balls!

Then the guard stopped smiling. He saw that the clown intended to cross the bridge.

"Wait!" he said. "How much do you weigh?"

"Why, exactly 198 pounds (89.10 kilograms) with my costume on," answered the clown.

The guard felt one of the colored balls.

"This weighs at least a pound (0.45 kg)," he said. "So you and the balls together weigh 201 pounds (90.45 kg)." He shook his head. "I'm sorry, but you can't cross the bridge."

"Why not?" asked the clown in wonder.

"Because the bridge will hold only 200 pounds (90 kg)," said the guard. "Any more than that and it will collapse. You would fall to your death in the deep canyon! That's why I'm here—to make sure that no one who's too heavy tries to cross the bridge."

"I've got to cross," said the clown. "My circus act starts in a few minutes! I must hurry!"

"Leave the balls behind."

The clown shook his head. "I can't. I need them for my act."

Suddenly, the clown smiled. "I know how I can do it," he exclaimed.

How did the clown cross the bridge safely?

(ANSWER ON PAGE 30)

The Ping-Pong and the Biggie

Explorers from earth landed on a far-off planet. They were amazed to find that all the creatures on the planet were shaped like balls! These creatures couldn't bounce, but they could move by rolling.

The smartest creatures were about the size of a ping-pong ball. The earth people called them "Ping-Pongs." There was another kind of creature that was ten times bigger than a Ping-Pong. The earth people called these things "Biggies." They noticed that anytime a Biggie saw a Ping-Pong, the Biggie would try to roll over the little Ping-Pong and crush it!

The earth people built a small, one-room hut to live in while they were on the planet. When they left the planet, they left this hut empty, with the door open.

One day, a Ping-Pong came rolling into the hut. It was amazed by the square room with the four sharp corners. Ping-Pongs had never imagined such a thing.

Suddenly, a Biggie came rolling into the hut. It was between the Ping-Pong and the door—the only way out. The Ping-Pong was trapped! The Biggie rolled toward it!

But the Ping-Pong was clever. It found a way to keep the Biggie from rolling over it. What did the Ping-Pong do?

(ANSWER ON PAGE 31)

Abdul's journeys

An Arab named Abdul made a journey from his home to another part of the desert.

For the first half of the journey, Abdul got to ride in a truck. It was a one-hour ride. And going by truck was ten times faster than Abdul could have walked the same distance.

However, for the second half of the journey, Abdul had to ride an old, lame camel. That part of the trip took him twice as long as if he had walked.

When Abdul went back home, he had to walk all the way. Which of the two trips was quickest? How much time did each trip take?

(ANSWER ON PAGE 31)

A mechanically minded squirrel

If you can figure out which way each gear is moving, you can tell whether the squirrel is pulling the basket of nuts *up* or letting it *down*. The first gear is turning in the direction of the dotted arrow. Here's a hint to get you started: the second gear will turn in the opposite direction.

(ANSWER ON PAGE 31)

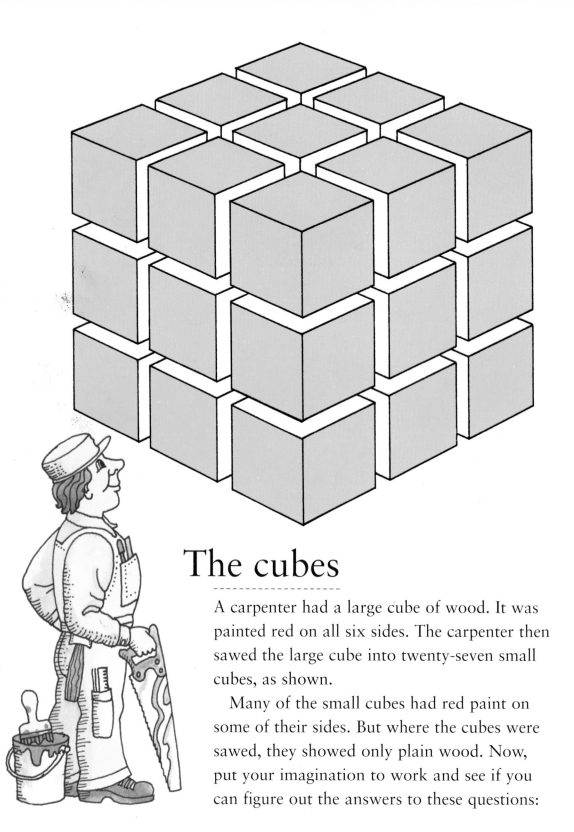

The cubes

A carpenter had a large cube of wood. It was painted red on all six sides. The carpenter then sawed the large cube into twenty-seven small cubes, as shown.

Many of the small cubes had red paint on some of their sides. But where the cubes were sawed, they showed only plain wood. Now, put your imagination to work and see if you can figure out the answers to these questions:

1. How many of the small cubes had red paint on three sides?

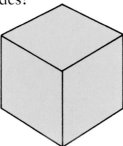

2. How many of the small cubes had red paint on two sides?

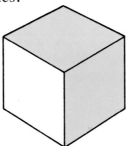

3. How many of the small cubes had red paint on only one side?

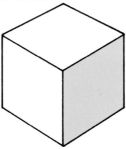

4. Were there any of the small cubes that had no paint on them at all?

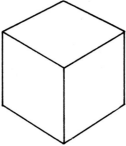

(ANSWERS ON PAGE 31)

Kliggs, klaggs, and kluggs

1. Let's call this shape a "kligg." This is one side of a kligg:

Now try to turn the kligg around in your mind. Which one of the drawings below shows the other side of a kligg?

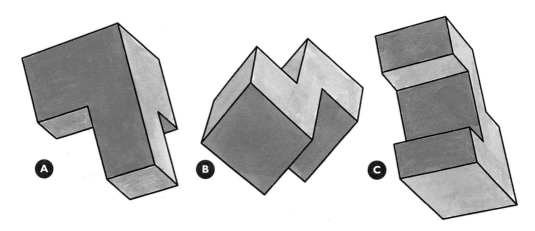

A   B   C

2. Let's call this shape a "klagg."
   This is one side of a klagg:

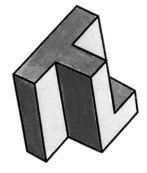

Now try to turn the klagg around
in your mind. Which one of the drawings
below shows the other side of a klagg?

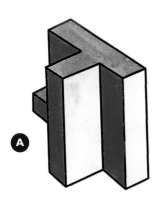

A

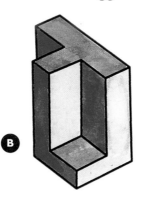

B

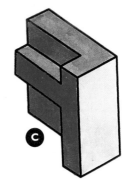

C

3. Let's call this shape a "klugg."
   This is one side of a klugg:

Now try to turn the klugg around in your
mind. Which one of the drawings below shows
the other side of a klugg?

(ANSWERS ON PAGE 32)

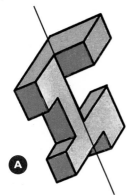

A

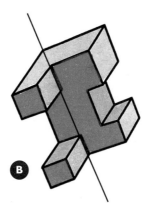

B

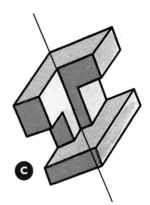

C

Word breakdown

To play this game, you start with a word, take away one letter, and make a new word out of the letters left. Then, take a letter out of that word, and make still another word with the letters that are left. Keep doing this until you have only a one-letter word—such as A or I—left.

For example, start with the word MAPLE. Take away the E and rearrange the M, A, P, and L to make the word PALM. Then, take away the L to make MAP. Next, drop the P and make AM. And, finally, take away the M and you're left with the word A.

Now, try it with these words.

ORANGE   STREAM

PARTED   RETAIN

NAILED   RELATE

(ANSWERS ON PAGE 32)

The bad bears

One day, Mrs. Bear went to market and left her four little cubs, Bruno, Boffin, Bobo, and Bilda, alone in the house. Of course, the four got into mischief! When Mrs. Bear returned, she found a broken honeypot and a big puddle of honey on the living room floor!

"Who broke the honeypot?" demanded Mrs. Bear.

"I didn't do it," cried Bruno.

"Bobo did it," claimed Boffin.

"Bilda did it," said Bobo.

"Bobo's lying!" exclaimed Bilda.

Three of the little bears are lying. One is telling the truth. Who is telling the truth? And which bear broke the honeypot?

(ANSWER ON PAGE 32)

The clever old woman

Long ago, in the days when people used candles to light their houses, there lived an old woman. She was poor, but very clever. She thought of many ways to save money.

One way she saved money was by making new candles from old ones. She found that one-fourth of a candle never burned because it was inside the candleholder. Most people just threw away these unburned candle stubs. But the old woman saved hers. By melting four of these stubs, she could make one new candle!

One day, the old woman went to the candlemaker's shop and bought sixteen candles. How many new candles could she make from the left-over fourths of the sixteen candles?

(ANSWER ON PAGE 32)

Answers

The people of Tuffleheim
(PAGE 7)

The traveler must ask a question to which he knows the answer. Then he will know whether the answer he got was true or false. He could ask a question with an obvious answer, such as "Is it daytime?" or "Am I a man?" A truth-teller would, of course, give the right answer, but a liar would reveal himself by giving an answer the traveler knows is wrong.

How many carnations? (PAGE 8)

The smallest number of carnations Wilfred McDoodle would have to bring back would be *three*. When he picked out two carnations, they might both be the same color, but one might be red and one might be green. Therefore, by taking along a third flower that would be either red or green, he would be sure of having two carnations of the same color.

The hungry rabbit (PAGE 9)

The rabbit must squeeze through the fence. She must push as many cabbages as she needs up to the fence. Then she can go back outside and eat the cabbages through the fence.

The outlaws and the marshals
(PAGE 10)

The sheriff knew that if the man who answered first had been an outlaw, he would have lied and said that he was a marshal. And if he were a marshal, he would have told the truth. So, whichever way, he would have said that he and the other untied man were marshals.

Therefore, when the other untied man told the sheriff that the first man had said they were marshals, he was obviously telling the truth. But if he had been an outlaw, he would have lied. Thus, the sheriff knew the two untied men were indeed the marshals, and the other two men were the outlaws.

Caspuccio the bandit (PAGE 12)

Caspuccio said, "You're going to cut off my head."

If the King said this was true, he'd have to behead Caspuccio to *make* it true. But he had said that the bandit would be hanged for making a true statement.

Answers

However, if Caspuccio were hanged, that would make his statement false—which would mean he should have been beheaded, making the statement true after all! So, the king couldn't decide whether the statement was true or false. He had to let Caspuccio go!

The magical beanies (PAGE 14)

Frank said he could see three red beanies and two white ones. But that's *five* beanies. And Frank can see only the beanies on the four other children, not his own. So he's lying, which means that he's wearing a red beanie.

Mark says he can see four white beanies. But we know that Frank is wearing a red beanie. Mark is lying when he says he sees only white beanies. So, Mark is also wearing a red beanie.

Kim says she sees one red and three white beanies. But both Frank and Mark have red ones. Kim is lying when she says she sees only one red beanie, so we know she's wearing a red beanie.

We know Frank, Mark, and Kim are wearing red beanies, so Hannah is telling the truth when she says she sees three red beanies. And we know that at least two of the children are wearing white beanies, so Hannah must be telling the truth when she says she sees one—otherwise, she would be wearing a red beanie and would have to lie about the white one.

John must be wearing the one white beanie Hannah sees, so he's a truth-teller, too. Thus, when he says he sees the same thing Hannah does, we know for sure that she has a white beanie on.

Frank, Mark, and Kim are wearing red beanies. Hannah and John are wearing white beanies.

The three musicians (PAGE 14)

First, the two children rowed to the other side of the river. One of them got out of the boat, and the other rowed back.

The first musician then rowed the boat across. When he reached the other side, the child who was there got into the boat and rowed it back.

Next, the two children took the boat across the river again. Again, one of them stayed there while the other brought the boat back. The second musician then rowed across.

This was done one more time, until all three musicians and one child were across the river. The musicians paid the child, who then rowed back across the river to his friend.

The brave clown (PAGE 16)

The clown juggled the balls as he walked across the bridge. This way, one of the balls was always in the air. Thus, the weight of the clown and the other two balls was exactly 200 pounds (90.45 kg), which the bridge could hold.

Answers

The Ping-Pong and the Biggie

(PAGE 18)

The Ping-Pong rolled into a corner. Because the Biggie was ten times bigger than the Ping-Pong, it could not fit into as small an angle as the Ping-Pong, as shown below. The Ping-Pong stayed in the corner until the Biggie grew tired and left.

Abdul's journeys (PAGE 20)

It was actually quicker for Abdul to walk than it was for him to go by truck and camel!

For the first journey, going half the distance by truck took one hour. This was ten times faster than Abdul could have walked. In other words, Abdul could have walked halfway in ten hours. Therefore, if his camel ride was twice as slow as walking, it must have taken twenty hours. So altogether, Abdul's first journey took twenty-one hours.

If Abdul could walk halfway in ten hours, it obviously took him twenty hours to walk the whole way. So his homeward journey, walking, took only twenty hours—one hour less than his first journey.

A mechanically minded squirrel

(PAGE 21)

The arrows show the direction in which each gear is moving. The last gear is moving counterclockwise—turning toward its left—so it is pulling the basket up.

The cubes (PAGE 22)

1. Eight cubes have red paint on three sides.
2. Twelve cubes have red paint on two sides.
3. Six cubes have red paint on one side.
4. Only one cube has no paint on it at all.

Answers

Kliggs, klaggs, and kluggs
(PAGE 24)

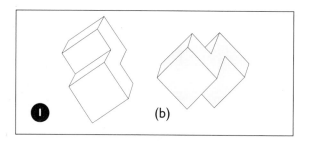

❶ (b)

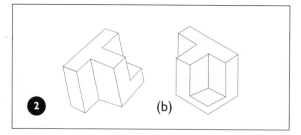

❷ (b)

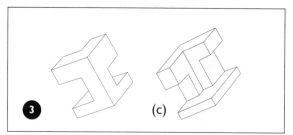

❸ (c)

The bad bears (PAGE 27)

If Bruno is telling the truth when he says that he didn't break the pot, then the other three statements are false. This means that Bobo is lying when he says Bilda broke the pot. However, this makes Bilda's statement, that Bobo is lying, true. We know that three of the bears are lying, so there can't be two true statements.

Therefore, Bruno's statement must be false, which means he broke the honeypot. Thus, Boffin's claim that Bobo broke it is false. So is Bobo's claim that Bilda broke it. Bilda's statement that Bobo is lying is the true one.

The clever old woman (PAGE 28)

The clever old woman made five new candles. Every four candles gave her four-fourths from which to make one new candle. So, from the sixteen candles she made four new ones. But, those four candles also gave her four-fourths from which she made a fifth candle.

Word breakdown (PAGE 26)

ORANGE	PARTED	NAILED	STREAM	RETAIN	RELATE
RANGE	RATED	LINED	MEATS	TRAIN	LATER
RANG	TEAR	DINE	SEAT	RAIN	LATE
RAN	TEA	DIN	SAT	RAN	ATE
AN	AT	IN	AT	AN	AT
A	A	I	A	A	A